BEI GRIN MACHT SICH IHR WISSEN BEZAHLT

- Wir veröffentlichen Ihre Hausarbeit,
  Bachelor- und Masterarbeit

- Ihr eigenes eBook und Buch -
  weltweit in allen wichtigen Shops

- Verdienen Sie an jedem Verkauf

Jetzt bei www.GRIN.com hochladen
und kostenlos publizieren

Impressum:

Copyright © 2016 GRIN Verlag, Open Publishing GmbH
Druck und Bindung: Books on Demand GmbH, Norderstedt Germany
ISBN: 9783656985686

Dieses Buch bei GRIN:

http://www.grin.com/de/e-book/334482/another-look-at-the-missing-mass-problem

Frank Armitage

Another look at the Missing Mass Problem

ABSTRACT

This paper questions the widely accepted concept of "dark matter", used to explain discrepancies between the current theory of gravity, as it relates to the visible mass of the universe (ie. stars, dust and gas), and the observations of the movements of stars in the spiral arms of galaxies and in clusters of galaxies. The idea of dark matter, developed in 1932 by Jan Oort and in 1933 by Fritz Zwicky, was used to explain why the observed movements of stars were are not following Kepler's third law, being the relationship between the distance of an orbiting object and it's orbital period. In this paper, I am suggesting that the gross effects of these observations are in reality the combined effects of both gravity and magnetic fields, and also the reasoning to support this hypothesis. In this paper, as well, I have tried to explain my hypothesis through language which is not overtly technical, so that an intelligent layperson would be able to understand it. It is hoped that the practical and theoretical significance of my paper will result in a re-examination of the basic, underlying foundations supporting the idea of dark matter, and that this re-examination will result in a theory which does not need to rely upon an invisible, untouchable, and all encompassing gravational source as being one of the basic planks of the theory of the universe.

Another look at the Missing Mass Problem   By Frank E. Armitage

Imagine, for a moment, two magnets in free fall. Over time, with nothing else affecting them, they will come together through both gravity and magnetism.

When they attempt to come together through gravity alone, the two objects are attracted to each other apparently through the average location of their mass distribution, that is where the force of gravity appears to be concentrated through the centre of gravity of each object.   The gross effect of this will appear to be as though the center of mass of each object will attract the other.

When they come together through both gravity and magnetism (for one cannot have a magnetic field without its source, which will itself generate gravity), as a direct result of the two magnetic fields, the two centers of mass will attempt to fall towards each other and, while this is occurring, the magnetic fields will attempt to re-orient themselves, that is to align each other, into a single North-South-North-South configuration.   That is that the north pole of the first magnet will be attracted towards the south pole of the second magnet.   As this is occurring, the magnetic field of both magnets together will be larger than that of either magnet alone.

When studying the cosmos, it has been noted that what seems to be a discrepancy exists with respect to the current theory of gravity and the distribution of stars in a spiral galaxy. This apparent discrepancy is that the rotation curve of a typical spiral galaxy arm does not follow what should be expected by gravitational theory alone [*1]. This is because, due to the large distances, the stars at the ends of the spiral arms should be travelling around the galaxy at a speed slower than those stars which are closer to the center of the galaxy. Yet, it has been found that the rotation speed of the stars at the ends of the spiral arms remains constant with, or even slightly faster than, those stars closer to the galaxy center [*2].   If most of the mass in the galaxy was concentrated in its center (ie. It's buldge) [*3], then this is contrary to Kepler's third law [*4].

The amount of attraction needed to allow these stars to remain where they are, and so to counteract the forces which should be causing them to simply slip away from the galaxy, clearly exceeds the gravitational attraction caused by the amount of mass visible in the galaxy itself [*5].

Since there must be a source for this additional force which keeps the arms of the spiral galaxy together [*6] - and assuming that the laws of gravity (and, of course, all of physics) are consistent throughout the universe [*7]- it has been proposed that there must exist some gravitational source to generate this force which has not, as yet, been observed.   Because of the presumption that the force needed must be gravity, and the fact that only mass can generate gravity, this apparent discrepancy was resolved by the proposition that there must exist an unobserved form of mass, which has been called "dark matter".

This model of dark matter, used to explain away the problem of too little observed mass, was first postulated by the Dutch astronomer, Jan Oort, in 1932, to account for the orbital velocities of stars in the Milky Way [*8].   In 1933, the Swiss astrophysicist, Fritz Zwicky, also used this concept to account for the missing mass in the orbital velocities of galaxies in clusters [*9].   This "missing mass problem" was that the gravity, which should be generated by what was observed to be in the galaxies in such clusters, was far too small to account for such fast orbits. Therefore, something else would be required to explain what held these orbits in place [*10].

This concept of dark matter was also used to explain other problems with what had been observed, such as gravitational lensing, various temperature distributions of hot gasses in galaxies and clusters of galaxies and, of course, the rotational speeds of galaxies [*11].

It has been calculated that the amount of dark matter needed to generate these additional gravitational effects would exceed, by several times, the amount of normal visible matter in the known universe [*12].

This "dark matter" seems to be an amazing thing. Not only can it not be seen, but it seems to be utterly transparent [*13]. It neither emits, nor has it any effect upon, light or any other type of electromagnetic radiation [*14]. It cannot be directly detected by either optical or radio telescopes. It cannot interact with ordinary matter through electromagnetic forces, and it does not carry any electric charge [*15]. Dispite these problems, as Georg Raffelt wrote in the Encyclopedia of Astronomy and Astrophysics, "The presence of large amounts of dark matter in the universe is almost uncontroversial." [*16].

Such a description reminds me of the theory of "luminiferous aether", which became, as stated in Wikipedia, "...so entrenched in physical law that it was simply assumed to exist.". [*17]. At the time, this theory, which was also used by Sir Isaac Newton when his corpuscle theory of light ran into difficulties [*18], was used as an explanation for the movement of light (and other electromagnetic energy) when it was shown that light behaves more as a wave than as a particle [*19].   This was because it was then believed that nothing wavelike could travel in a vacuum [*20].   Therefore, the thinking went, there had to be some medium through which these waves could travel between the sun, other bodies and the earth.   This medium, which was believed to have physical properties in itself [*21], was "luminiferous aether" [*22].   The main problem with this theory, however, was what exactly was this luminiferous aether?

As more and better theories and observational discoveries resulted in larger and greater difficulties with the theory, the medium of luminious aether was forced to evolve into something that became awkward to describe:

> "It had to be a fluid in order to fill spaces, but one that was millions of times more rigid than steel in order to support the high frequencies of light waves.   It also had to be massless and without viscosity, otherwise it would visibly affect the orbits of planets.   Additionally, it appeared it had to be completely transparent, non-dispersive, incompressible, and continuous at a very small scale." [*23].

This entrenched belief in aether continued, despite such major difficulties in the theory, until the theories of special relativity and quantum mechanics established that it was simply no longer needed [*24].   Only then was the theory determined to be incorrect.

The concept of "dark matter" is inferred to exist only because of the apparent gravitational effects on visible matter [*25].   Basically, the idea behind dark matter was to account for the gravitational discrepency between the masses of galaxies, clusters of galaxies, and larger structures, which had been determined through the relativistic calculations of the masses of the perceived (ie. visible) matter of these objects - being the stars, dust and gasses found (and estimated) in and around these structures – and the fact that the gravational formulas, which work so well in our solar syatem, don't appear to work for the orbits of some of the stars in these same structures.   Thus, dark matter is simply a classical solution to a classical gravity problem, an example of Occam's Razor used to arrive at a simple means to account for and describe what is otherwise inexplicable [*26].

Thus, the evidence for dark matter is inferred only through the observed motions of galaxies [*27], and its presence is presumed only by what appears to be gravitational attraction [*28].

An example of this thinking, by using gravity in it's common usage, would have an object, "A", resting on the surface of a larger object, "B", and being unable to get away from "B" without the use of a force sufficient to push (or pull) "A" away from "B".   "A" would be held onto "B" by gravity. If we were only talking about gravity, then the only way "A" would be able to leave "B" would be for "A" to be drawn towards an object with a larger mass, "C". Because of the larger force of attraction between "A" and "C", than exists between "A" and "B", "A" would be drawn towards "C", and so would "fall" off of "B" towards "C" until it was resting on the surface of "C". For this example, "B" would also be falling towards "C", but it would be doing this by moving at a slower speed than "A" (thanks to inertia).   This is because the force needed to overcome inertia for a large object is greater than that needed for a smaller object. Thus, although both "A" and "B" are being affected by "C"'s gravitational attraction, "A" would move toward "C" faster than would "B".

Such a simplistic example follows the basic parameters of gravity, but also ignores the fact that in the real universe there is more than simply gravity. Let us change the above example to make "A2" a magnet with the same mass as "A", and also add "D", something a magnet will be drawn to with a mass far smaller than either "B" or "C". If you place "D" near "A2", "A2" will be drawn to "D", and so will "fall" off of the surface of "B" towards "D", until it is resting on the surface of "D". In the short distances we are talking about, the magnetic attraction between "A2" and "D" is stronger than the gravitational attraction between either "A2" and "B" or "A2" and "C", and so "A2" will continue to rest on the surface of "D" notwithstanding the gravitational attraction towards "A2" from both "B" and "C". Note that "D" may also fall from "B" to "C", but that "A2" will still remain attracted to, and so continue being on the surface of, "D".

Ignoring, for the moment, what forces we are talking about, the net observable effect of the position of "A2", being on the surface of "D", and of "A2" resisting the need to fall to either "B" or "C", would appear, to someone unaware of which forces we are talking about, that the gravitational attraction between "A2" and "D" is greater than that between "B" and/or "C". Because it is clear that the mass of "D" is less than the mass of either "B" and "C", this would result in an apparent discrepancy in the law of gravity. However, in reality, there would be no such discrepancy, as the primary force of attraction between "A2" and "D" is magnetic, rather than through gravity alone, and so "A2" would remain on the surface of "D" despite the larger gravitational attraction to "B" or "C".

Such an effect would be an example of the symptomology of gravity, that is that a small object ("A") will remain on the surface of a larger object ("B") unless and until it comes across something with a larger attraction ("C"), to move the small object from the smaller to the larger object (that is from "B" to "C"). The same crude, net effect occurs with magnetism, in that a small magnetic object ("A2") will remain on the surface of another magnetic object ("D") unless and until it comes across something with a higher magnetic attraction ("E"), which is able to move "A2" from "D" to "E". Thus, this effect of magnetism appears to be gravity-like, except that the sizes (read masses) of the objects are not the only relevant moving force. Rather, in the case of "A2", the primary moving force is magnetism.

In the case of magnetism, however, both "D" and "E" can be smaller in mass than "B" and "C", and we all know that a magnetic field can overwhelm gravitational attraction. There is nothing new here at all.

Over interstellar distances, however, the view is that it is gravity which is the primary force of movement between stars. This is because magnetic effects, though powerful up close, are considered to be overwhelmed by distance. A magnetic field can be described as a dipole field characterized by its total magnetic dipole moment [*29]. One characteristic of a dipole field is that the strength of the field falls off inversely with the cube of the distance from the magnet's centre [*30]. Gravity only falls off with the square of the distance [*31].

Thus, although it is clear that the Milky Way's magnetic field is directed along the spiral arms [*32], the popular view is that such forces will not have any major effect upon the positioning of stars in the galaxy [*33].

Now although the strength of the dipole field will fall off inversely with the cube of the distance from the magnet's center, the strength of the magnetic field would be determined by the strength of the magnetic field generator itself.   Stars are massive and powerful magnetic field generators.   As Dr. Mark Miesch [*34] says in his paper, "Solar internal flows and dynamo action", the "stars bristle with magnetic energy" [*35]. Indeed, he says that as much as 0.1% of our own suns luminosity may be converted to magnetic energy just by the sun's solar convection zone [*36].

Using the example of our sun, which is itself a magnetic field generator, it's diameter is about 1.393 million kilometres [*37], while it's magnetic field has a diameter of several light hours [*38].   It must also be remembered that a magnetic field is not just a single point.   As with mass, the location of the apparent source of the magnetic field will be the average location of the distribution of the sources of each individual magnetic dipole moment, being from a number of diffuse points in space which surrounds the magnetic source itself.   Each of these points will be interacting with each other, but will also be interacting with the larger and more powerful initial source.

The size and strength of the magnetic field will be generated by the strength of the magnetic source, and this magnetic field will exist far outside of the confines of the physical volume of the magnetic source [*39].   The source of gravity, on the other hand, ends where the mass itself ends.

A magnetic field is naturally oriented to its magnetic source.   That is the North Pole of the magnetic field always wants to be oriented towards the North Pole of the magnetic source.   [This can be shown by holding a magnet close to a magnetic compass and moving it around.   The needle of the compass will follow the magnet around, showing the location of the north, or south, pole of the dominant magnetic field of the magnet.]

This is because the magnetic field exists only because of the magnetic source itself. And it is the magnetic field, generated by the magnetic source, which causes the gross effect seen in magnetism: that is that one magnet field is attracted to, or repulsed from, another .   The source of the magnetic field, and the magnetic field itself, will influence its absolute three dimensional orientation in the universe, while trying to maintain its relative positioning regarding its various Poles.

Each star - and it's magnetic field - can be visualized as a sphere within a sphere (if you would), each of which can twirl (that is orient itself) in all three dimensions.   The physical volume of the star's magnetic field is far larger than the size of the star itself. Normally, the magnetic field wants to stay oriented with the star's magnetic poles.   The orientation of the star's magnetic field is dependent upon the orientation of the magnetic poles of the star, in that the north magnetic dipole moment of the star's magnetic field will be alligned with the magnetic north pole of the star. Conversely, the star's magnetic poles also want to stay oriented with it's magnetic field.

Because of this, in a weightless, or zero-g, environment, if a force encourages the movement of the star's magnetic field to become oriented in a way which is not alligned to the star itself, the star itself will attempt to realign itself and, thus, re-establish the normal, natural alignment between the star and it's magnetic field. True the mass of the star will not want to do this (thanks to Newton's first law and inertia), but, in a zero-g environment where there is a constant force upon the mass of the star, over time such a re-orientation of the star will occur.

Now, imagine a universe containing only two magnets, both of which are separated by distance "d". Regardless of the distance, both magnets will have an effect upon the other, from being either overwhelming to essentially zero plus something small [*40]. This same sort of effect exists for gravitational attraction, in that bodies may be separated by a distance of millions of light years, resulting in an almost, but not quite, zero attraction to each other.

When these magnetic fields comes into contact with each other, they will both want to change each other's orientations from where they are, vis a vis their initial three dimensional orientation (which is dependent upon the three dimensional orientation of their magnetic sources), and re-orient themselves towards the natural North-South-North-South positioning of magnets. They are both held back from doing this by the inertia resulting from the mass of each magnetic source which will resist such movement. Thus, each individual magnetic source will continue to direct that the north pole of its own magnetic field shall continue to be aligned with the north pole of itself, while the magnetic fields will be exerting a force upon each other to attempt to re-orient themselves into the basic North-South-North-South configuration.

In essence, the force on each magnetic field will want to drag its particular magnetic source towards where the magnetic field wants it to go. But, to be able to do so this, the magnetic field will need to overcome the inertia of the mass of the magnetic source.   (For this example, I am ignoring the problem of friction, as my hypothesis is based upon magnetic sources in micro-gravity, ie. "free fall". However, the same effects would occur when overcoming friction).

In a micro-gravity environment, a small amount of force (such as the magnetic dipole - dipole interaction between the magnetic fields set out above) will, over time, be able to cause significant variations in the three dimensional orientation of both magnets. This is because the force acting upon the two magnetic fields will slowly re-orient the two magnetic sources towards the North-South-North-South configuration which is preferred by the two magnetic fields.

A similar example of a small force having a major effect over time can be seen with the idea of solar sails. Using the radiation pressure of light, such sails can be used as a form of propulsion, to push objects in the vacuum of outer space. Although this pressure is very weak, once you are away from the drag caused by gravity, an atmosphere or some other source of friction, there will be nothing to slow down or otherwise hinder the acceleration resulting from these sails. As such, over time, the speed generated by these solar sails can build up and become quite significant [*41].

Remember, in micro-gravity, there would be nothing to hinder or interfere with the force between the two magnetic fields noted above save and except for their inertia (and, of course, their gravitational attraction, which should not have any real effect upon this orientation as the gravitational attraction would be affecting the mass of the two sources through their center of mass). Thus the magnets, being pulled by their magnetic fields towards a particular orientation, would, over a sufficient amount of time, move towards the orientation favoured by the magnetic fields.

When this occurs, the two magnetic sources will be oriented into a North-South-North-South configuration, and the total magnetic field, formed by the two magnetic sources, will be larger than either of the two individual magnetic fields acting alone.

As noted above, however, the real universe contains more than simply two magnets. Every star and each of the other charged particles in the universe are magnetic sources, too [*42]. In addition, most of these bodies are in what is effectively micro-gravity, or "free fall". Such a state can certainly exist while in the orbit of a galaxy (so long as an object follows the orbital model required by gravity, it can be oriented in any direction due to micro-gravity - just like being "weightless" while orbiting the earth). As such, over time, they would each want to become oriented into a magnetic arrangement whereby a number of them would be oriented into their own North-South-North-South formation.

It must be remembered that while the source of gravity is concentrated in relatively small volumes of space, the magnetic fields generated by a magnetic source encompass a far larger volume of space. It is suggested that the larger volume of space which contains the magnetic field holds a gross magnetic moment which, although comparatively weak relative to a set volume of space, is sufficient throughout the entirety of the field to have a combined major influence over other magnetic fields in free fall over a substantial distance .

Now, at first blush, it would appear that a problem exists with my hypothesis in that the sun, a normal magnetically active star, regularly reverses it's magnetic fields - that is the Sun's magnetic north and south poles reverse direction - every nine to twelve years during it's solar maximum [*43]. As such, one must also conclude that the same effects will occur on most, if not all, other stars. Yet, my hypothesis rests not on the transitory orientation of the magnetic source, involving a particular direction of the magnetic poles, but in the fact that the magnetic field itself exists.

This is because of the gross physical volume of the magnetic fields affecting each other. It is suggested that the total force from each of these various magnetic fields should be able to, over time and under the effect of micro-gravity, have a substantial and material effect upon other magnetic fields as to their location, which should also affect the locations of their magnetic sources. As with gravity, where the sun, though far away from the star Proxima Centauri [*44], still has a gravitational effect upon it, the substantially larger physical volume of the magnetic field of the sun would also have an effect upon the magnetic field of Proxima Centauri. This effect of this would be to move their three dimensional orientations to one favoured by their magnetic fields.

Again, assuming the universe contains only two magnets, when the magnetic field from the first magnet impacts upon the magnetic field from the second magnet, their effects are not merely to attract or repel the other, but are intended to reorient both magnetic fields into a combined North-South-North-South orientation. Such an effect would be based upon the strength of each of the magnetic sources which would affect the volume and strength of the magnetic fields.   Remember, a magnetic source does not generate the magnetic field merely from the center of the source. The center of the magnetic source, like the center of mass, is just a single point from which all of the magnetic generating points are averaged out. This is because a Ferromagnetic material [*45] is composed of the alignment of magnetic moments associated with electrons in the atomic lattice [*46]. These magnetic moments assemble themselves into spontaniously magnetized groupings called "domains" which, under the influence of an external magnetic field, can orient themselves into the direction of the magnetic field [*47].

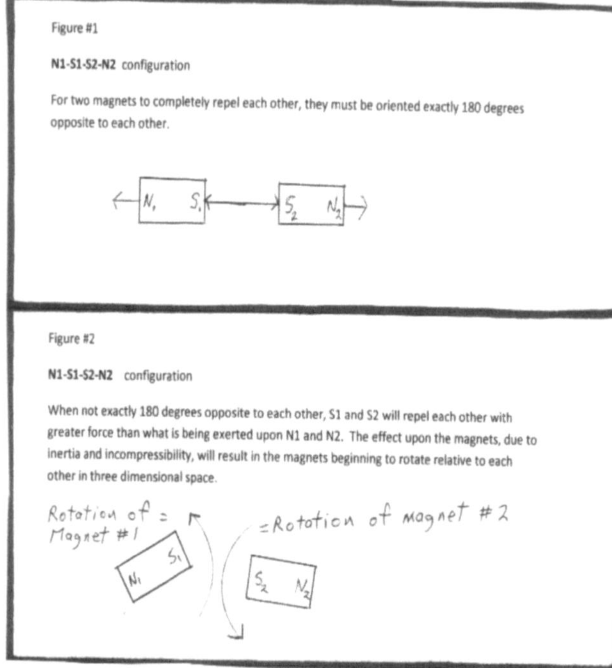

Each of these magnetic generating points is its own magnetic source and many additional magnetic sources can be created by being exposed to a magnetic field [*48]. Using the example of our sun, it's magnetic field extends out to a volume being several light hours from the physical surface of the sun [*49].

It is also interesting to note that approximately 90% of the atoms of the known universe is composed of hydrogen [*50], and hydrogen, as a dilute monatomic ion, is paramagnetic [*51]. Indeed, every material that is exposed to a magnetic field becomes, in various ways and strengths, magnetic itself [*52]. The strength of the magnetic moment of these materials from the exposure of the magnetic field depends upon both the material itself as well as the strength of the magnetic field.

Using the example of magnets again, generally speaking, when two magnets are in a precise North-South-South-North configuration, which is exactly 180 degrees opposite to each other [see Figure 1], the magnetic field will simply repel both magnets.

However, when the same orientation of North-South-South-North exists, but where the magnets are not exactly 180 degrees opposite to each other [see Figure 2], the repulsive force acting on the poles which are closer to each other will be greater than on the poles which are further away from each other.

This force will therefore push the poles that are closer to each other apart faster than the poles which are further apart. Because of inertia and the incompressibility of the magnetic sources, the result of this would be that the magnetic fields will force the two magnetic sources to rotate around their individual centers of mass in three dimensional space, with the North pole of one magnetic source moving towards the South pole of the second magnetic source. This rotation will also be assisted, although weakly, by the North pole from one magnet increasing its attraction to the South pole of the second magnet - as they get nearer to each other - while they rotate around their Centers of Mass.

Since the three dimensional orientation of the magnetic source is the primary determining factor of the three dimensional orientation of the magnetic field, from this one may conclude that one magnetic field will always attract a second magnetic field except where these two magnetic fields are oriented exactly 180 degrees opposite to each other.

This will apply only if there are two magnets (or magnetic sources) in the universe. In reality, there are numerous magnetic sources throughout the universe, and so their combined magnetic fields will ensure that no single magnetic source will be in an orientation which is exactly 180 degrees opposite to that of another magnetic source for longer than a brief amount of time.

 Thus, when the poles of a magnetic source reverse themselves, the fact that the poles may either attract or repel the poles of the other magnet will only effect the poles, not the magnetic fields themselves since both magnetic fields are interacting with and so, in effect, attracting the other.

As the actual volume of magnetic fields can be extremely large, I suggest that the consequence of such fields will have a surprisingly large effect upon the surrounding magnetic fields. I suggest, as well, that the effect of magnetic fields upon the orientation and positioning of other magnetic fields - and, as a consequence, the sources of the magnetic fields - throughout the galaxy may very well exceed the gravitational attraction between the masses of the sources of the magnetic fields themselves.

The theory of dark matter assumes that what surrounds and encompasses a galaxy is what is called a halo of dark matter. An assumption for this "dark matter halo" model is that the halo is spherical [*53], and has a density distribution which is approximately that of an isothermal sphere [*54], extending far beyond the galaxy and dominating its gravitational field in the outer regions. If true, then the stars just outside the outer limits of a galaxy should also be spinning with (or just slightly slower than) the galaxy itself. Thus, it should be possible to test my hypothesis by determining the velocity of those stars which are near to, but not in, a galactic spiral arm. These stars would be just above or below the disk of the galaxy and, if dark matter does not exist, these stars should be more or less following Kepler's Third Law. I suggest that this may be tested using the instruments used by Rubin, Ford jr. and Thonnard to observe the outer limits of a galaxy [*55].

Another look at the
Missing Mass Problem   Frank E. Armitage

It is proposed that the combined gravitational and magnetic effects as discussed with this paper, both working in tandem, may be able to assist in answering the "missing mass problem" set out by Oort and Zwicky.

Footnotes:

[*1]   Foundations of Modern Cosmology (2nd ed.), pages 439 - 440.

[*2]   Dark Side of the Universe, page 36.

[*3]   Concepts of Modern Physics: The Haifa Lectures, page 96; Dark Matter: Its Nature, page. 584.

[*4]   Kepler's Third Law shows the relationship between the distance of a planet from the sun and its orbital period. This formula is used for any object orbiting another object. (see "Kepler's Laws", Encyclopedia of Astronomy and Astrophysics, p. 1368; and Foundations of Modern Cosmology, pps. 439 - 440.).

[*5]   Dark Side of the Universe, page 36.

[*6]   Rotational Properties of 21 Sc Galaxies with a Large Range of Luminosities and Radii, from NGC 4605 (R = 4 kpc) to UGC 2885 (R = 122 kpc), page 485.

[*7]   Which also presumes that modified Newtonian dynamics is incorrect, see Dark Matter: Its Nature, page 589.

[*8]   The Whole Shebang, page 123.

[*9]   "Dark Matter: Its Nature", Encyclopedia of Astronomy and Astrophysics, p. 584.

[*10]   "Dark Matter: Its Nature", Encyclopedia of Astronomy and Astrophysics, p. 583.

[*11]   Dark Side of the Universe, pages 44 – 47.

[*12]   "Dark Matter in Galaxies", Encyclopedia of Astronomy and Astrophysics, p. 582.

[*13]   Dark Side of the Universe, page 33.

[*14]   Dark Side of the Universe, page 33.

[*15]   "Dark matter", page 3, Wikipedia.

[*16]   "Dark Matter: Its Nature", Encyclopedia of Astronomy and Astrophysics, p. 587.

[*17]   "Luminiferous aether", page 4, Wikipedia.

[*18]   North Pole South Pole, page 95.

[*19]   North Pole South Pole, page 95.

[*20]   "Ether", Encyclopedia Britannica (9th ed.), Vol. 8, pages 568 – 572.

[*21]   Magnetic Fields: A Comprehensive Theoritical Treatise for Practical Use, page 125.

[*22]   Also spelled "Ether", A Dictionary of Astronomy, page 154.

[*23]   "Luminiferous aether", page 4, Wikipedia.

[*24]   "ether", A Dictionary of Astronomy, page 154.

[*25]   "Dark Matter: Its Nature", Encyclopedia of Astronomy and Astrophysics .

[*26]   "(F)or any two competing interpretations (or theories), the one with the least assumptions is the
         better one.", see "Occam's razor", Dictionary of Geophysics, Astrogeophysics and Astronomy,
         Page 342.

[*27]   "Dark Matter in Galaxies", ", Encyclopedia of Astronomy and Astrophysics .

[*28]   Dark Side of the Universe, page 37.

[*29]   "Force between magnets", page 2, Wikipedia.

[*30]   "force between magnets", page 2, Wikipedia.

[*31]   "The Law of Universal Gravitation", Cambridge Encyclopedia of Astronomy, page 400.

[*32]   "What do we really know about the magnetic fields of the Milky Way?" page 515, Cosmic
         Magnetic Fields: From Planets, to Stars and Galaxies.

[*33]   Although there are still serious questions arising regarding what role the magnetic field plays in
         the interstellar medium of irregular galaxies and spiral galaxies (see "Magnetic fields in irregular
         galaxies", Cosmic Magnetic Fields: From Planets, to Stars and Galaxies).

[*34]   Dr. Mark Miesch received his PhD in Astrophysics, and currently works at the High Altitude
         Observatory of the U.S. National Center for Atmospheric Research.

[*35]   "Solar internal flows and dynamo action", page 99.

[*36]   "Solar internal flows and dynamo action", page 100.

[*37]   The Sun, page 15.

[*38]   Physics of the Sun: A First Course, page 335.

[*39]   Magnetic Fields: A Comprehensive Theoritical Treatise for Practical Use, page 125.

[*40]   <u>Magnetic Fields: A Comprehensive Theoretical Treatise for Practical Use</u>, page 125.

[*41]   <u>Solar Sails</u>, chapter 6.

[*42]   <u>Physics of Ferromagnetism</u>, page 7.

[*43]   <u>Physics of the Sun: A First Course</u>, page 277.

[*44]   The closest star to the sun, being 4.22 light years away (see "Proxima Centauri", <u>Encyclopedia of Astronomy and Astrophysics</u>, page 2171).

[*45]   <u>Physics of Ferromagnetism</u>, page 118.

[*46]   <u>Permanent Magnets and Their Application</u>, page 50.

[*47]   <u>Permanent Magnets and Their Application</u>, page 50.

[*48]   <u>Introduction to the Theory of Ferromagnetism</u>, page 1.

[*49]   <u>Physics of the Sun: A First Course</u>, page 335.

[*50]   <u>Hydrogen: The Essential Element</u>, page 9.

[*51]   <u>Physics of Ferromagnetism</u>, Table 4.1, page 85.

[*52]   <u>Introduction to the Theory of Ferromagnetism</u>, page 1.

[*53]   As a working model for the theory of dark matter, it is merely an assumption that this dark matter halo is spherical. Since no one has observed this halo, according to the mathematics it could be ellipsoid in shape. For the purposes of testing my hypothesis, however, its supposed shape is irrelevant (<u>Dark Side of the Universe</u>, page 36).

[*54]   See Begeman, K.G.; Broeils, A.H.; Sanders, R.H.; "Extended rotation curves of spiral galaxies: dark haloes and modified dynamics", <u>Monthly Notices of the Royal Astronomical Society</u>, vol. 249, April 1, 1991, p. 523, at 526.

[*55]   They were able to examine the rotational properties of spiral Sc galaxies to their faint outer limits, (see Rubin, V.C.; Ford jr, W.K.; Thonnard, N.; "Rotational Properties of 21 Sc Galaxies with a Large Range of Luminosities and Radii, from NCG 4605 (R= 4 kpc) to UGC 2885 (R=122 kpc)", <u>Astrophysical Journal</u>, vol. 238, p. 471.)

BIBLIOGRAPHY

Aharoni, A., Introduction to the Theory of Ferromagnetism, Oxford University Press, (1996).

Begeman, K.G., Broeils, A.H., Sanders, R.H., "Extended rotation curves of spiral galaxies: dark haloes and modified dynamics.", Monthly Notices of the Royal Astronomical Society, vol. 249, pps. 523 - 537.

Chown, M., "Ditching dark matter", The Guardian (Newspaper), Thursday, February 13, 2003.

Concepts of Modern Physics: The Haifa Lectures, Lecture IX, "The Universe", Imperial College Press (2007).

Dictionary of Geophysics, Astrogeophysics and Astronomy, Matzner, R.A. (Ed.), CRC Press LLC, (2001).

Encyclopedia of Astronomy and Astrophysics, IOP Publishing Ltd. and Nature Publishing Group (2001).

"Ether", Encyclopedia Britannica (9th ed.), Vol. #8, New Youk: Samuel Hall (1878).

Ferris, T., The Whole Shebang: A State-of-the-Universe(s) Report, Simon & Schuster (1997).

Hawley, J.F., Holcomb, K.A., Foundations of Modern Cosmology (2nd ed.), Oxford University Press (2005).

Kepley, A., Wilcots, E., Zweibel, E., Muhle, S., Everett, J., Robishaw, T., Heiles, C., Klein, U., "Magnetic fields in irregular galaxies", K.G.Strassmeier, A.G.Kosovichev, J.E.Beckman, eds., Cosmic Magnetic Fields: From Planets to Stars and Galaxies, Cambridge University Press (2009).

Miesch, M.S., "Solar internal flows and dynamo action", Heliophysics: Evolving Solar Activity and the Climates of Space and Earth, C.J. Schrijver and G.L. Siscoe (eds.), Cambridge University Press (2010).

Mitton, S. (ed.), The Cambridge Encyclopedia of Astronomy, Trewin Copplestone Publishing Ltd, (1977).

Mullan, D.J., Physics of the Sun: A First Course, Chapman & Hall/CRC (2010).

Nicholson, I., The Sun, Mitchell Beazley Publishers (1982).

Parker, R.J., Studders, R.J., Permanent Magnets and Their Application, John Wiley & Sons, Inc., (1962).

Purrington, R.D., Physics in the Nineteenth Century, Rutgers University Press, (1997).

Rigden, J.S., Hydrogen: The Essential Element, Harvard University Press, (2002).

Rubin, V., Thonnard, N., Ford, jr., W.K., "Rotational Properties of 21 Sc Galaxies with a Large Range of Luminosities and Radii from NGC 4605 (R = 4 kpc) to UGC 2885 (R = 122 kpc)", Astrophysical Journal, vol. 238, p.471.

Turner, G., North Pole, South Pole: The Epic Quest to Solve the Great Mystery of Earth's Magnetism, The Experiment, LLC,(2011).

Trimble, V., "Existence and nature of dark matter in the universe", Annual Review of Astronomy and Astrophysics, Vol. #25, pages 425 – 472.

Vulpetti, G., Johnson, L., Matloff, G.L., Solar Sails: A Novel Approach to Interplanetory Travel, Praxis Publishing Ltd., (2008).

Wielebinski, R., "What do we really know about the magnetic fields of the Milky Way?", K.G.Strassmeier, A.G.Kosovichev, J.E.Beckman, eds., Cosmic Magnetic Fields: From Planets to Stars and Galaxies, Cambridge University Press (2009).

Wikipedia, the free encyclopedia,   http://en.wikipedia.org
- Center of mass
- Centrifugal force
- Dark matter
- Dark matter halo
- Diamagnetism
- Ferromagnetism
- Force between magnets
- Galaxy rotation curve
- Gravitation
- Hydrogen
- Intermolecular force
- Kepler's laws of planetary motion
- Luminiferous aether
- Magnetic dipole-dipole interaction
- Magnetic field
- Magnetic moment
- Modified Newtonian dynamics
- Paramagnetism
- Sun, the
- van der Waals forces

Zirker,J.B., The Magnetic Universe: The Elusive Traces of an Invisible Force, The Johns Hopkins University Press, (2009).